The Sustainable Future A Blueprint for Habitat Conservation and Restoration

Sylvia

TABLE OF CONTENTS

Chapter 1: Introduction to Ecological Restoration and Habitat Conservation — 06

The Importance of Ecological Restoration

The Significance of Habitat Conservation

Chapter 2: Understanding Ecosystems and Biodiversity — 10

The Basics of Ecosystems

Exploring Biodiversity and its Importance

Chapter 3: Threats to Habitat Conservation — 14

Deforestation and Its Impacts

Urbanization and Loss of Natural Habitats

Climate Change and its Effects on Habitats

Chapter 4: Principles of Ecological Restoration — 20

Defining Ecological Restoration

The Three R's of Restoration: Repair, Reconnect, and Reestablish

Assessing Site Conditions for Restoration Projects

Chapter 5: Techniques for Habitat Restoration 26

Native Plant Restoration

Wetland Restoration

Forest Restoration

Marine and Coastal Habitat Restoration

Chapter 6: Conservation Strategies for Habitat Protection 34

Protected Areas and National Parks

Land Trusts and Conservation Easements

Sustainable Agriculture and Wildlife Conservation

Chapter 7: Community Engagement and Stakeholder Involvement 41

The Role of Local Communities in Conservation Efforts

Collaborating with Government Agencies and NGOs

Educating and Empowering Individuals for Sustainable Action

Chapter 8: Case Studies in Successful Habitat Conservation and Restoration 47

Restoring Degraded Farmlands into Thriving Wildlife Habitats

Urban Greening Initiatives: Transforming Cities into Biodiversity Hotspots

Mangrove Restoration: Protecting Coastal Communities and Marine Life

Chapter 9: Challenges and Future Directions 53

Overcoming Financial and Political Barriers

Integrating Conservation and Restoration Efforts on a Global Scale

Embracing Technology and Innovation for Effective Habitat Management

Chapter 10: The Role of Every Individual in Creating a Sustainable Future 59

Practical Steps for Promoting Habitat Conservation in Daily Life

Inspiring Others to Join the Movement

Building a Network of Sustainable Communities

Conclusion: A Call to Action for a Sustainable Future 65

Chapter 1: Introduction to Ecological Restoration and Habitat Conservation

The Importance of Ecological Restoration

In a rapidly changing world, the need for ecological restoration has become increasingly critical. The delicate balance of our ecosystems is being threatened by human activities such as deforestation, pollution, and climate change. The consequences of these actions are far-reaching, affecting not only the environment but also our own health and well-being. Therefore, it is essential for us to understand the importance of ecological restoration and take action to preserve and restore our natural habitats.

Ecological restoration is the process of repairing and rehabilitating ecosystems that have been degraded or destroyed. It involves a wide range of activities, including reforestation, habitat creation, and the removal of pollutants from soil and water. By restoring ecosystems, we can help to reverse the damage caused by human activities and promote the recovery of biodiversity.

One of the key reasons why ecological restoration is important is its role in preserving biodiversity. Ecosystems are home to a wide variety of plant and animal species, all of which are interconnected. When an ecosystem is degraded, the loss of habitat and resources can lead to the decline or extinction of species. By restoring ecosystems, we can provide a safe and thriving habitat for these species, ensuring their survival for future generations.

Ecological restoration also plays a crucial role in mitigating the impacts of climate change. Healthy ecosystems act as carbon sinks, absorbing and storing large amounts of carbon dioxide from the atmosphere. By restoring degraded ecosystems, we can enhance their capacity to sequester carbon, helping to reduce greenhouse gas emissions and combat climate change.

Furthermore, ecological restoration can provide numerous benefits to human communities. Restored ecosystems improve water quality, reduce the risk of floods, and provide a source of clean air. They also offer recreational opportunities, promote tourism, and contribute to the overall well-being of communities. By investing in ecological restoration, we can create sustainable and resilient habitats that support both wildlife and human populations.

In conclusion, the importance of ecological restoration cannot be overstated. It is a vital tool in conserving our environment, protecting biodiversity, mitigating climate change, and improving the quality of life for all. As individuals, we have a responsibility to support and engage in restoration efforts, whether through volunteering, advocating for policy changes, or practicing sustainable living. By working together, we can create a sustainable future for ourselves and future generations.

The Significance of Habitat Conservation

Habitat conservation plays a vital role in ensuring the sustainable future of our planet. In the face of rapid urbanization, industrialization, and climate change, it is essential to protect and restore natural habitats for the benefit of all living beings. This subchapter explores the significance of habitat conservation and its relevance to the field of environmental engineering.

First and foremost, habitat conservation is crucial for preserving biodiversity. Habitats provide a home to countless species, each playing a unique role in maintaining the delicate balance of ecosystems. By conserving habitats, we can safeguard the rich tapestry of life on Earth and prevent the extinction of species. Environmental engineers play a key role in this process by designing and implementing strategies to protect and restore habitats, ensuring the survival of both endangered species and their habitats.

Furthermore, habitat conservation contributes to the overall health and well-being of our planet. Natural habitats act as carbon sinks, absorbing carbon dioxide from the atmosphere and mitigating the impacts of climate change. They also provide essential ecosystem services such as water purification, flood control, and soil stabilization. Environmental engineers are at the forefront of developing innovative solutions to restore degraded habitats, enhancing their capacity to absorb carbon, filter water, and provide other critical services.

Habitat conservation also has significant economic benefits. Many industries rely on natural resources obtained from habitats, such as timber, medicinal plants, and clean water. By conserving habitats, we ensure the sustainable availability of these resources, supporting industries and providing livelihoods for local communities.

Environmental engineers play a crucial role in finding sustainable ways to extract resources from habitats without causing irreversible damage or depletion.

Moreover, habitat conservation is closely linked to human health and well-being. Access to green spaces and natural habitats has been proven to have numerous physical and mental health benefits. Conserving habitats in and around urban areas promotes a healthier and more sustainable lifestyle for communities. Environmental engineers can contribute by designing urban green spaces that mimic natural habitats, providing residents with opportunities to connect with nature and improve their overall well-being.

In conclusion, habitat conservation is of utmost significance in securing a sustainable future for all. It ensures the preservation of biodiversity, contributes to climate change mitigation, provides essential ecosystem services, supports economic development, and promotes human health and well-being. Environmental engineers have a crucial role to play in the protection and restoration of habitats, employing their expertise to design and implement innovative solutions. By recognizing the importance of habitat conservation and actively participating in its implementation, we can pave the way for a sustainable future for generations to come.

Chapter 2: Understanding Ecosystems and Biodiversity

The Basics of Ecosystems

Ecosystems are the building blocks of our planet's intricate web of life. They encompass all living organisms and their physical surroundings, functioning as interconnected systems where each component plays a vital role. Understanding the basics of ecosystems is crucial for anyone interested in environmental engineering and the sustainable future of our planet.

At its core, an ecosystem consists of two main components: biotic and abiotic. Biotic factors refer to all living organisms, from the tiniest microbes to the largest mammals, including plants, animals, fungi, and bacteria. On the other hand, abiotic factors encompass all non-living elements, such as air, water, sunlight, soil, and climate. The interaction between these biotic and abiotic factors is what shapes the dynamics of ecosystems.

Ecosystems vary greatly in size and complexity, ranging from a small pond to vast rainforests or even the entire planet. Regardless of their scale, they all follow the same fundamental principles. The first principle is that all elements within an ecosystem are interconnected, creating a delicate balance. A disturbance or change in one component can have cascading effects throughout the entire system.

The second principle is the concept of energy flow. Energy enters ecosystems primarily through sunlight, which is transformed by plants during photosynthesis into chemical energy. This energy is then transferred from one organism to another as they consume and are consumed, forming food chains and webs. Decomposers play a vital role in breaking down organic matter, returning nutrients to the environment and closing the loop of energy flow.

The third principle is the cycling of nutrients. Nutrients, such as carbon, nitrogen, and phosphorus, are essential for the growth and survival of organisms. They move through the ecosystem in a continuous cycle, being taken up by plants, consumed by animals, and eventually released back into the environment through decomposition. This recycling process ensures the sustainability of ecosystems.

Human activities have a profound impact on ecosystems, often leading to habitat destruction, pollution, and the loss of biodiversity. As environmental engineers, we have a responsibility to mitigate these negative effects and work towards habitat conservation and restoration. By understanding the basics of ecosystems, we can develop innovative solutions that promote sustainability and ensure the well-being of both present and future generations.

In conclusion, ecosystems are the foundation of life on Earth, encompassing all living organisms and their physical surroundings. They operate on the principles of interconnectedness, energy flow, and nutrient cycling. As environmental engineers, it is our duty to protect and restore ecosystems, safeguarding our planet's biodiversity and creating a sustainable future for all.

Exploring Biodiversity and its Importance

Biodiversity, or the variety of life on Earth, is a fundamental pillar of our planet's health and sustenance. In this subchapter, we will delve into the concept of biodiversity, why it is crucial for our environment, and its significance for addressing the challenges faced by environmental engineers.

Biodiversity encompasses all living organisms, including plants, animals, and microorganisms, and the intricate web of interactions they form within their ecosystems. It is the foundation of thriving ecosystems, providing essential services such as clean air and water, nutrient cycling, and climate regulation. Without biodiversity, our natural systems would collapse, leading to severe consequences for all life forms, including humans.

Understanding the importance of biodiversity is crucial for environmental engineers as they play a vital role in designing sustainable solutions to address the environmental challenges we face today. By comprehending the intricate relationships between different species and their habitats, engineers can develop innovative approaches to habitat conservation and restoration.

One significant aspect of biodiversity is its role in supporting the resilience of ecosystems. Diverse ecosystems are better equipped to withstand disturbances, such as climate change or natural disasters, as the presence of a variety of species allows for greater adaptation and recovery. Environmental engineers can leverage this knowledge to design and implement restoration projects that enhance the biodiversity of degraded habitats, ensuring their long-term survival and functionality.

Furthermore, biodiversity holds immense potential for addressing pressing environmental issues, such as pollution and resource depletion. Many species possess unique traits and genetic material that can be harnessed for the development of sustainable technologies and solutions. By studying and preserving biodiversity, environmental engineers can unlock nature's secrets and develop innovative approaches to waste management, renewable energy, and sustainable agriculture.

In conclusion, exploring biodiversity and understanding its importance is crucial for environmental engineers. By recognizing the intricate connections between species and their habitats, engineers can develop sustainable solutions for habitat conservation and restoration. Biodiversity is not only essential for the health of ecosystems but also holds great potential for addressing environmental challenges through the development of sustainable technologies. As we move towards a sustainable future, it is imperative that we value and protect the rich tapestry of life on Earth and harness its potential for the betterment of all.

Chapter 3: Threats to Habitat Conservation

Deforestation and Its Impacts

Deforestation is an issue of critical importance in today's world, as it not only affects the environment but also has far-reaching social and economic consequences. In this subchapter, we will explore the causes and impacts of deforestation, as well as potential solutions to mitigate its effects.

Deforestation refers to the permanent removal of forests or woodland areas, typically for the purpose of converting the land for agriculture, logging, or urban development. It is estimated that approximately 18 million acres of forest are lost each year, contributing to the degradation of natural habitats and the loss of biodiversity.

One of the main drivers behind deforestation is the demand for agricultural land. As the global population continues to grow, there is an increasing need for food production, leading to the conversion of forests into farmland. This not only results in the loss of diverse plant and animal species but also disrupts the delicate balance of ecosystems.

The impacts of deforestation are wide-ranging and profound. Firstly, it contributes to climate change by releasing large amounts of carbon dioxide into the atmosphere. Forests act as carbon sinks, absorbing carbon dioxide and helping to regulate the Earth's climate. When trees are cut down, this stored carbon is released, contributing to the greenhouse effect and global warming.

Additionally, deforestation disrupts water cycles and can lead to soil erosion. Forests play a crucial role in maintaining water balance by absorbing and storing rainfall. When forests are cleared, water runoff increases, leading to flooding in some areas and drought in others.

Moreover, the removal of trees exposes the soil to erosion, further degrading the land and reducing its productivity.

The impacts of deforestation extend beyond the environment and affect human communities as well. Indigenous peoples and local communities who depend on forests for their livelihoods are often disproportionately affected, as deforestation destroys their homes and traditional ways of life.

To address the issue of deforestation, it is essential to implement sustainable land management practices and promote reforestation efforts. This includes the establishment of protected areas, the adoption of sustainable agricultural practices, and the promotion of responsible logging practices. Additionally, raising awareness and educating the public about the importance of forests and the consequences of deforestation is crucial.

In conclusion, deforestation poses significant challenges to environmental engineering and sustainable development. By understanding the causes and impacts of deforestation, we can work towards finding innovative solutions to protect and restore our forests, ensuring a sustainable future for all.

Urbanization and Loss of Natural Habitats

In recent decades, urbanization has emerged as a global phenomenon, transforming landscapes and altering ecosystems at an unprecedented rate. As the world's population continues to grow and cities expand, the loss of natural habitats has become a pressing issue that demands immediate attention. This subchapter delves into the adverse effects of urbanization on natural habitats and highlights the importance of sustainable practices in environmental engineering.

Urbanization, characterized by the rapid growth of cities and infrastructure development, often results in the destruction or fragmentation of natural habitats. As urban areas expand, forests, wetlands, and grasslands are cleared to make way for buildings, roads, and other urban infrastructure. This widespread habitat loss has detrimental consequences for both wildlife and the overall ecological balance.

The loss of natural habitats disrupts critical ecosystems, threatening the survival of numerous plant and animal species. Urbanization fragments habitats, creating isolated patches that are unable to sustain viable populations. This fragmentation limits gene flow, increases the risk of inbreeding, and reduces biodiversity. Furthermore, as natural habitats shrink, wildlife is forced into closer contact with human settlements, leading to conflicts and increased risks of disease transmission.

Environmental engineering plays a crucial role in mitigating the negative impacts of urbanization on natural habitats. By employing sustainable practices, engineers can design and develop cities that minimize habitat loss and preserve biodiversity. This involves incorporating green spaces, such as parks and urban forests, into urban planning. These green spaces serve as valuable refuges for wildlife and help to maintain ecological connectivity between fragments of natural habitats.

Furthermore, environmental engineers can promote the use of green infrastructure, such as green roofs and permeable pavements, which mimic natural processes and reduce the impervious surfaces that contribute to habitat loss. Implementing sustainable stormwater management systems that mimic natural hydrological processes can also help to preserve wetland habitats and reduce the risk of flooding.

In conclusion, urbanization poses significant challenges to the preservation of natural habitats. However, through the application of sustainable practices in environmental engineering, we can minimize the loss of habitats and create cities that coexist harmoniously with nature. By prioritizing biodiversity conservation and adopting green infrastructure solutions, we can ensure a sustainable future where urbanization and habitat preservation go hand in hand. It is imperative for all individuals, especially those in the field of environmental engineering, to work collectively to protect and restore our natural habitats for the benefit of present and future generations.

Climate Change and its Effects on Habitats

Climate change is a pressing issue that affects every living being on this planet. It is a result of human-induced activities, primarily the burning of fossil fuels, which release greenhouse gases into the atmosphere, trapping heat and causing a rise in global temperatures. These rising temperatures have far-reaching consequences, particularly on the habitats that support our planet's diverse ecosystems.

In the subchapter, "Climate Change and its Effects on Habitats," we will explore how environmental engineering can play a vital role in understanding and mitigating these effects for the sustainable future of our planet.

One of the most significant effects of climate change on habitats is the alteration of temperature and precipitation patterns. Rising temperatures can disrupt the natural balance of ecosystems, leading to the loss of biodiversity and the extinction of vulnerable species. Environmental engineers can analyze these changes and propose innovative strategies to restore and conserve these habitats.

Another consequence of climate change is the melting of polar ice caps and glaciers, resulting in rising sea levels. This phenomenon poses a significant threat to coastal habitats and communities. Environmental engineers can design and implement coastal protection measures to mitigate the impact of rising sea levels, such as constructing seawalls, restoring natural shorelines, or implementing sustainable coastal development practices.

Additionally, climate change has led to more frequent and intense extreme weather events like hurricanes, floods, and droughts. These events can further damage habitats, disrupt ecosystems, and threaten the livelihoods of communities. Environmental engineers can develop sustainable infrastructure and implement nature-based solutions to reduce the vulnerability of habitats and communities to these disasters.

Furthermore, climate change has implications for freshwater habitats, affecting the availability and quality of water resources. Environmental engineers can develop innovative technologies for water conservation and purification, ensuring the sustainability of freshwater habitats and meeting the growing demands of human populations.

In conclusion, "Climate Change and its Effects on Habitats" highlights the critical role of environmental engineering in understanding, mitigating, and adapting to the consequences of climate change. By implementing sustainable strategies and developing innovative technologies, we can conserve and restore habitats, safeguard biodiversity, and ensure a sustainable future for all. This subchapter aims to inspire and empower environmental engineering professionals to take action and contribute to the blueprint for habitat conservation and restoration.

Chapter 4: Principles of Ecological Restoration

Defining Ecological Restoration

Ecological restoration is a fundamental concept in the field of environmental engineering, and it plays a crucial role in creating a sustainable future for our planet. In this subchapter, we will delve into the meaning and importance of ecological restoration, exploring its relevance to habitat conservation and its potential benefits for all individuals.

At its core, ecological restoration refers to the process of renewing, rehabilitating, or restoring ecosystems that have been degraded, damaged, or destroyed. It involves a variety of activities aimed at bringing back the natural balance and functionality of an ecosystem, such as reintroducing native species, restoring water bodies, and managing invasive species. The ultimate goal is to recreate a self-sustaining and resilient ecosystem that can provide essential ecological services and support biodiversity.

The need for ecological restoration arises from the recognition that human activities have significantly impacted the environment, leading to the loss of biodiversity, degradation of natural habitats, and disruption of ecosystem functions. Environmental engineering professionals have a critical role to play in reversing these trends and restoring the health and functionality of ecosystems.

Ecological restoration offers numerous benefits, not only for the environment but also for society as a whole. By restoring degraded habitats, we can enhance biodiversity, which is crucial for the stability and resilience of ecosystems. Healthy ecosystems provide various services, such as clean air and water, soil fertility, and climate regulation, which are essential for human well-being and survival.

Furthermore, ecological restoration can contribute to the mitigation of climate change. Restored ecosystems can sequester carbon dioxide, reducing greenhouse gas levels in the atmosphere. They can also serve as buffers against natural disasters, such as floods and storms, by providing natural floodplains and coastal defenses.

One of the key aspects of ecological restoration is the involvement of stakeholders and the wider community. It is not only the responsibility of environmental engineers but also requires the active participation and support of everyone. By understanding the importance of ecological restoration and actively engaging in restoration efforts, we can all contribute to a sustainable future.

In conclusion, ecological restoration is a vital concept in the field of environmental engineering. It entails the process of restoring degraded ecosystems to their natural state, providing numerous benefits for the environment and society as a whole. By actively participating in ecological restoration efforts, we can contribute to the preservation of biodiversity, mitigation of climate change, and the creation of a sustainable future for all.

The Three R's of Restoration: Repair, Reconnect, and Reestablish

Title: The Three R's of Restoration: Repair, Reconnect, and Reestablish

Introduction:

In the quest for a sustainable future, habitat conservation and restoration play a crucial role. As environmental engineers, we have the power to make a positive impact on our planet. The Three R's of Restoration - Repair, Reconnect, and Reestablish - provide a blueprint for achieving effective and long-lasting habitat restoration efforts. By understanding and implementing these principles, we can pave the way for a healthier and more resilient ecosystem for future generations.

Repair:

To repair a degraded habitat, we must first identify and address the root causes of its decline. This may involve rehabilitating polluted water sources, remediating soil contamination, or restoring native vegetation. Repairing the damage done to our ecosystems requires a multi-disciplinary approach, combining scientific research, engineering expertise, and community engagement. By repairing the ecological infrastructure, we can restore the natural balance and enhance the biodiversity that sustains our planet.

Reconnect:

Habitats are interconnected networks, and their fragmentation hinders the movement of species and disrupts ecological processes. Reconnecting fragmented habitats is essential for restoring balance and resilience to ecosystems. This can be achieved through the creation of wildlife corridors, green belts, or the removal of physical barriers such as dams or roads. By fostering connectivity, we enable the free flow of genetic material, increase biodiversity, and enhance the overall health and vitality of our ecosystems.

Reestablish:

Restoration is not just about fixing what is broken; it is also about reestablishing what has been lost. Many species have been pushed to the brink of extinction, and their absence has severe consequences for the ecological balance. Reestablishing these species in their natural habitats through reintroduction programs is crucial for maintaining healthy ecosystems. By working closely with conservation organizations, we can facilitate the reestablishment of keystone species, ensuring their survival and the recovery of their respective habitats.

Conclusion:

As environmental engineers, it is our responsibility to be at the forefront of habitat conservation and restoration efforts. The Three R's of Restoration - Repair, Reconnect, and Reestablish - provide a comprehensive framework for achieving sustainable and effective restoration outcomes. By repairing damaged ecosystems, reconnecting fragmented habitats, and reestablishing vital species, we can create a sustainable future for all. Let us embrace these principles and work collectively towards a planet where humans and nature coexist in harmony. Together, we can make a profound difference and leave a lasting legacy for generations to come.

Assessing Site Conditions for Restoration Projects

In the realm of habitat conservation and restoration, evaluating site conditions is a crucial step that sets the foundation for successful environmental engineering projects. This subchapter will delve into the importance of assessing site conditions and provide a comprehensive guide for professionals and enthusiasts alike.

The first step towards effective restoration is understanding the current state of the site. This involves evaluating the soil composition, topography, and hydrology, among other factors. By conducting a thorough assessment, environmental engineers can identify potential challenges and develop appropriate strategies to address them. For instance, if the soil is contaminated or lacks essential nutrients, remediation efforts can be tailored accordingly to ensure the successful establishment of native vegetation.

Hydrological conditions play a significant role in restoration projects as they influence the availability of water and determine the overall health of the ecosystem. Assessing site conditions involves examining the water flow, drainage patterns, and water quality. This information helps engineers design effective water management systems, such as the installation of erosion control measures or the implementation of stormwater management techniques.

Additionally, understanding the existing vegetation and wildlife on the site is crucial to preserving biodiversity and promoting ecological balance. Assessing the flora and fauna present provides valuable insights into the site's ecological functions and potential restoration opportunities. It allows engineers to identify key species and habitats that need to be protected or restored, ensuring that the project aligns with conservation goals.

Furthermore, assessing site conditions involves considering the potential impacts of climate change. Environmental engineers need to evaluate the site's vulnerability to rising temperatures, changing precipitation patterns, or sea-level rise. By anticipating these impacts, they can develop adaptive management strategies and incorporate climate-resilient features into restoration plans.

The information gathered during site assessments serves as a basis for developing restoration objectives and designing appropriate restoration strategies. It allows environmental engineers to create tailored plans that consider the unique conditions of each site, maximizing the chances of success and long-term sustainability.

In conclusion, assessing site conditions is a critical step in any restoration project. By evaluating the soil, hydrological conditions, existing vegetation, and potential climate change impacts, environmental engineers can develop effective strategies to conserve and restore habitats. This subchapter provides an essential guide for professionals in the field of environmental engineering, ensuring that restoration efforts are well-informed, impactful, and sustainable. Whether you are a seasoned environmental engineer or a curious enthusiast, understanding how to assess site conditions is key to contributing to a sustainable future.

Chapter 5: Techniques for Habitat Restoration

Native Plant Restoration

Native plant restoration plays a crucial role in preserving and restoring the natural habitat, and it is an essential component of environmental engineering. This subchapter aims to provide an understanding of the importance of native plant restoration, its benefits, and the methods used to restore native plant communities.

Native plants are those that have evolved in a specific region over thousands of years, adapting to the local climate, soil, and wildlife. These plants provide numerous benefits to the environment, including soil stabilization, erosion control, water filtration, and the creation of wildlife habitat. However, due to urbanization, agriculture, and invasive species, many native plant communities have been degraded or completely lost.

Restoring native plant communities is vital for several reasons. Firstly, native plants are better adapted to local conditions, requiring less water, fertilizer, and pesticide use compared to non-native species, making them more sustainable and cost-effective. Secondly, native plants provide food and habitat for a variety of insects, birds, and other wildlife, supporting biodiversity and ecological balance. Finally, native plants help mitigate the impacts of climate change by sequestering carbon dioxide and reducing the urban heat island effect.

There are various methods used in native plant restoration, depending on the specific site conditions and goals. One common approach is the removal of invasive species that outcompete native plants for resources. This is often followed by the reintroduction or seeding of native plant species that were historically present in the area. In some cases, habitat restoration may involve the physical manipulation of the landscape, such as regrading or restoring wetlands.

Successful native plant restoration requires careful planning, site assessment, and ongoing management. It is essential to consider factors such as soil type, moisture levels, and sunlight availability when selecting appropriate native plant species. Additionally, monitoring and maintenance are crucial to ensure the long-term success of restoration efforts, including controlling invasive species, providing supplemental irrigation, and managing competing vegetation.

In conclusion, native plant restoration is a vital component of environmental engineering, contributing to the conservation and restoration of natural habitats. By restoring native plant communities, we can promote biodiversity, enhance ecosystem services, and create sustainable landscapes. Everyone has a role to play in supporting native plant restoration, whether through volunteering, advocating for conservation policies, or incorporating native plants into their own landscapes. Together, we can work towards a sustainable future that preserves and restores our precious natural heritage.

Wetland Restoration

Wetlands are among the most valuable ecosystems on our planet, providing a wide range of ecological services. Unfortunately, human activities have led to the degradation and destruction of many wetland areas worldwide. In order to mitigate the negative impacts of these actions and preserve the unique biodiversity and functionality of wetlands, the concept of wetland restoration has emerged as a crucial strategy.

Wetland restoration refers to the deliberate process of reestablishing the natural balance and functioning of wetland ecosystems that have been degraded or lost. This process involves the implementation of various techniques and approaches to recreate or enhance wetland habitats, including the reestablishment of hydrological conditions, the removal of invasive species, and the reintroduction of native flora and fauna.

One of the key motivations behind wetland restoration is the recognition of the vital role that wetlands play in maintaining environmental health. Wetlands act as natural filters, purifying water by trapping sediments and removing pollutants. They also provide flood control by absorbing excess water during heavy rainfall or storms, reducing the risk of downstream flooding.

Furthermore, wetlands are incredibly diverse ecosystems, supporting a wide array of plant and animal species. They serve as important breeding grounds and habitats for migratory birds, fish, amphibians, and countless other wildlife. Restoring wetlands not only helps protect these species but also ensures the preservation of their natural habitats and the ecological interactions that they depend on.

For environmental engineers, wetland restoration presents a unique challenge and opportunity. These professionals are at the forefront of designing and implementing restoration projects, utilizing their expertise in hydrology, ecology, and engineering principles. By understanding the intricate dynamics of wetland systems, environmental engineers can develop effective strategies to restore and maintain the ecological balance of these habitats.

Wetland restoration is not only beneficial for the environment but also for communities. Restored wetlands can provide recreational opportunities, such as hiking, fishing, and birdwatching, attracting tourists and boosting local economies. Additionally, the preservation of wetlands can enhance water quality and quantity, ensuring a sustainable supply of clean drinking water for nearby communities.

In conclusion, wetland restoration is a critical tool in the conservation and restoration of these valuable ecosystems. By implementing restoration practices, we can revive degraded wetlands, protect biodiversity, and maintain the ecological services that wetlands provide. Environmental engineers play a crucial role in this process, utilizing their skills and knowledge to design and implement effective restoration projects. The restoration of wetlands not only benefits the environment but also provides numerous social and economic advantages for communities and future generations.

Forest Restoration

Forests are essential for maintaining a sustainable future for our planet. They provide numerous ecosystem services, including carbon sequestration, biodiversity conservation, water regulation, and soil protection. However, due to human activities such as deforestation, forest degradation, and climate change, our forests are facing unprecedented challenges. To ensure the survival of these crucial ecosystems, forest restoration has become an urgent priority.

Forest restoration refers to the process of rebuilding, rehabilitating, or creating forests in areas that have been previously deforested or degraded. It involves a range of activities, including tree planting, natural regeneration, and the implementation of sustainable land management practices. Forest restoration is a multifaceted approach that not only focuses on increasing tree cover but also aims to restore the ecological integrity and functionality of the forest ecosystem.

Environmental engineering plays a crucial role in forest restoration efforts. Through the application of engineering principles, techniques, and technologies, environmental engineers contribute to the planning, design, and implementation of effective forest restoration projects. They assess the ecological and hydrological conditions of degraded forests, develop restoration plans, and monitor the progress of restoration activities.

One key aspect of forest restoration is reforestation. Environmental engineers work closely with ecologists and foresters to identify suitable tree species that are adapted to local conditions and promote biodiversity. They also consider factors such as soil composition, climate, and hydrology to ensure the successful establishment and growth of trees. Additionally, they may employ innovative techniques such as aerial seeding, direct seeding, or the use of seedlings to maximize the efficiency and effectiveness of reforestation efforts.

Another important aspect of forest restoration is the restoration of degraded ecosystems. Environmental engineers play a vital role in rehabilitating forest ecosystems that have been severely impacted by human activities, such as mining, logging, or agricultural expansion. They develop strategies to remediate contaminated soils, restore water bodies, and reintroduce native vegetation. By applying their expertise in soil and water management, environmental engineers help to restore the ecological balance of degraded forests and promote their resilience to future disturbances.

Forest restoration is not a one-size-fits-all solution. It requires a collaborative effort from various stakeholders, including governments, local communities, NGOs, and researchers. Environmental engineers serve as facilitators, bringing together different perspectives and expertise to develop comprehensive restoration plans that are socially acceptable, economically viable, and ecologically sound.

In conclusion, forest restoration is a critical component of habitat conservation and plays a vital role in ensuring a sustainable future for our planet. Environmental engineers are at the forefront of these efforts, using their knowledge and skills to restore degraded forests, enhance biodiversity, and mitigate climate change. By restoring our forests, we can safeguard the invaluable benefits they provide to both humans and the environment. It is a collective responsibility of every individual to support and engage in forest restoration initiatives to secure a sustainable future for generations to come.

Marine and Coastal Habitat Restoration

In recent years, there has been growing concern about the state of our marine and coastal ecosystems. These areas, which are home to a diverse range of plant and animal species, play a crucial role in maintaining the health of our planet. However, human activities such as pollution, overfishing, and coastal development have had a devastating impact on these fragile habitats. The need for marine and coastal habitat restoration has never been more urgent.

Restoring marine and coastal habitats involves a range of techniques aimed at improving the health and resilience of these ecosystems. One common approach is the creation of artificial reefs. These structures, made from materials such as concrete or sunken ships, provide a solid substrate for marine organisms to attach to and create new habitats. Artificial reefs not only provide shelter for fish and other marine species but also help to protect coastlines from erosion.

Another important aspect of marine and coastal habitat restoration is the removal of marine debris. Plastic pollution has become a global crisis, with millions of tons of plastic waste entering our oceans each year. This debris poses a significant threat to marine life, as animals can become entangled in it or mistake it for food. By organizing beach clean-ups and implementing strategies to reduce plastic waste, we can help restore the health of our marine and coastal ecosystems.

Coastal wetlands, such as mangroves and salt marshes, are also vital habitats that require restoration. These areas provide essential breeding grounds and nurseries for a wide variety of marine species. They act as natural filters, purifying water and protecting coastlines from storm surges. Unfortunately, coastal wetlands are disappearing at an alarming rate due to human activities. Restoring and conserving these habitats is crucial for maintaining the delicate balance of our coastal ecosystems.

Environmental engineering plays a crucial role in marine and coastal habitat restoration. Engineers work to design and implement sustainable solutions that can help restore and protect these vital ecosystems. They develop innovative technologies for cleaning up oil spills, designing coastal protection measures, and implementing sustainable fishing practices. Collaboration between environmental engineers, scientists, and policymakers is key to successfully restoring and conserving our marine and coastal habitats.

In conclusion, marine and coastal habitat restoration is of utmost importance in ensuring the health and sustainability of our planet. By implementing strategies such as creating artificial reefs, removing marine debris, and restoring coastal wetlands, we can help recover these valuable ecosystems. Environmental engineering plays a vital role in developing sustainable solutions for habitat restoration. It is crucial that we all come together to protect and restore our marine and coastal habitats for the benefit of current and future generations.

Chapter 6: Conservation Strategies for Habitat Protection

Protected Areas and National Parks

Introduction:

Protected areas and national parks play a crucial role in habitat conservation and restoration efforts. These designated areas aim to protect and preserve natural ecosystems, biodiversity, and cultural heritage for present and future generations. This subchapter explores the significance of these areas and highlights the role of environmental engineering in ensuring their sustainability.

Importance of Protected Areas: Protected areas are essential for maintaining ecological balance and safeguarding vulnerable species. These areas act as sanctuaries for endangered plants and animals, providing them with a secure habitat and promoting their conservation. They also serve as crucial corridors for wildlife to migrate and maintain genetic diversity. Additionally, protected areas offer recreational opportunities, education, and research possibilities, contributing to the overall well-being of societies.

Role of National Parks: National parks are a subset of protected areas, managed to preserve exceptional natural and cultural features. These parks often serve as flagship sites for conservation efforts, attracting visitors and raising awareness about environmental issues. National parks can also contribute to local economies through ecotourism, while simultaneously protecting fragile ecosystems. By designating and effectively managing national parks, we can ensure the long-term survival of unique landscapes and biodiversity hotspots.

Environmental Engineering in Protected Areas: Environmental engineering plays a pivotal role in the sustainable management of protected areas and national parks. These professionals utilize scientific knowledge and engineering techniques to address various environmental challenges. They design and implement infrastructure to support visitor management, including trails, campsites, and visitor centers, while minimizing the environmental impact. Environmental engineers also develop strategies for waste management, water conservation, and renewable energy use within these areas.

Furthermore, environmental engineers contribute to maintaining the ecological health of protected areas by monitoring and managing water quality, air pollution, and soil erosion. They work closely with park authorities to assess the carrying capacity of these areas and develop strategies for sustainable development. By integrating engineering principles with conservation goals, environmental engineers ensure that protected areas continue to thrive while accommodating human activities.

Conclusion:
Protected areas and national parks are invaluable assets to our planet. By preserving these areas and their biodiversity, we can secure a sustainable future for all species. Environmental engineering plays a crucial role in maintaining the delicate balance between conservation and human needs. By embracing the principles of sustainable development, we can ensure the long-term viability of protected areas

and national parks, allowing future generations to enjoy the wonders of nature while fostering environmental stewardship.

Land Trusts and Conservation Easements

In the pursuit of a sustainable future, it is essential to explore innovative approaches to habitat conservation and restoration. Land trusts and conservation easements have emerged as powerful tools in this endeavor, empowering individuals and organizations to protect and preserve critical natural resources for future generations. This subchapter delves into the concept of land trusts and conservation easements, their importance, and their role in environmental engineering.

Land trusts are nonprofit organizations dedicated to conserving land and natural resources. They work closely with landowners, communities, and government agencies to acquire and manage land for conservation purposes. By holding and managing land, land trusts ensure its long-term protection and responsible use. These organizations play a crucial role in safeguarding ecosystems, protecting biodiversity, and preserving important habitats.

One of the key mechanisms employed by land trusts is the conservation easement. A conservation easement is a legal agreement between a landowner and a land trust, which restricts certain activities on the land to protect its conservation values. These easements can be tailored to specific needs, allowing landowners to retain ownership while ensuring the land's ecological integrity. Conservation easements can address a range of conservation objectives, such as protecting endangered species, preserving water quality, or maintaining scenic landscapes.

Environmental engineering professionals play a vital role in the success of land trusts and conservation easements. Their expertise in assessing ecological systems, understanding environmental regulations, and implementing sustainable practices is invaluable in guiding land conservation efforts. Environmental engineers can collaborate with land trusts to conduct environmental assessments, develop conservation plans, and implement restoration projects. By leveraging their technical skills, these professionals can help ensure the long-term viability and effectiveness of land trusts and conservation easements.

For everyone, land trusts and conservation easements offer numerous benefits. They protect natural areas, safeguard wildlife habitats, and maintain the integrity of ecosystems. By conserving land, these initiatives also contribute to combating climate change, as forests and wetlands act as carbon sinks, sequestering greenhouse gases. Additionally, land trusts often provide public access to protected areas, allowing individuals to connect with nature, engage in recreational activities, and learn about the importance of conservation.

In conclusion, land trusts and conservation easements are crucial components of a sustainable future. By partnering with landowners, communities, and environmental engineering professionals, these initiatives ensure the long-term protection and responsible use of land and natural resources. Their impact extends beyond environmental preservation, contributing to climate change mitigation, public engagement, and the overall well-being of communities. Embracing and supporting land trusts and conservation easements is a collective responsibility that will shape a brighter and more sustainable future for all.

Sustainable Agriculture and Wildlife Conservation

In this subchapter, we explore the intricate relationship between sustainable agriculture and wildlife conservation, highlighting the critical role that environmental engineering plays in ensuring a harmonious coexistence between human activities and the natural world. As our planet faces numerous challenges, including climate change, biodiversity loss, and food security concerns, it is essential to adopt sustainable practices that protect both agricultural productivity and ecological integrity.

Sustainable agriculture refers to a system of farming that aims to produce food while minimizing environmental damage, preserving natural resources, and supporting rural communities. By implementing techniques such as agroforestry, organic farming, and integrated pest management, sustainable agriculture promotes biodiversity conservation, reduces soil erosion, and mitigates the use of harmful chemicals. These practices create a healthier and more resilient ecosystem, facilitating the coexistence of agriculture and wildlife.

Environmental engineering plays a crucial role in developing and implementing sustainable agriculture practices. Environmental engineers utilize their expertise to design and optimize irrigation systems, develop efficient waste management strategies, and design eco-friendly infrastructure that minimizes the ecological footprint of agricultural activities. They also contribute to the development of innovative technologies that enhance productivity while reducing environmental impact, such as precision agriculture and hydroponics.

The conservation of wildlife within agricultural landscapes is of paramount importance. By preserving natural habitats and creating wildlife corridors, farmers and environmental engineers can promote the survival of various species, including pollinators, birds, and beneficial insects. These organisms play a fundamental role in maintaining ecosystem balance, ensuring successful crop pollination, and providing natural pest control.

Furthermore, sustainable agriculture practices can help mitigate climate change by sequestering carbon in the soil, reducing greenhouse gas emissions, and enhancing overall resilience to extreme weather events. By employing techniques such as cover cropping, crop rotation, and agroforestry, farmers can sequester atmospheric carbon dioxide and mitigate its impact on global warming.

In conclusion, the integration of sustainable agriculture and wildlife conservation is crucial for the long-term well-being of our planet. Environmental engineering provides the knowledge and tools necessary to design and implement sustainable practices that protect both agricultural productivity and biodiversity. By adopting these practices, we can ensure a sustainable future, where agriculture thrives, wildlife flourishes, and humanity coexists harmoniously with nature. This subchapter serves as a blueprint for environmental engineers, farmers, policymakers, and all individuals interested in creating a more sustainable and resilient world.

Chapter 7: Community Engagement and Stakeholder Involvement

The Role of Local Communities in Conservation Efforts

In our pursuit of a sustainable future, it is crucial to recognize the integral role that local communities play in conservation efforts. As we strive for habitat conservation and restoration, it is important to understand how each individual, regardless of their background or expertise, can contribute to protecting our environment. This subchapter will delve into the significance of local communities in conservation and highlight the ways in which they can actively participate in preserving our natural resources.

Environmental engineering professionals, as well as anyone interested in habitat conservation, should recognize the immense potential that lies within local communities. These communities are uniquely positioned to understand and address the specific conservation needs of their surroundings. With their firsthand knowledge of the environment, they can identify critical ecosystems, endangered species, and areas of concern that require immediate attention.

Furthermore, local communities possess a deep connection and attachment to their surroundings. By engaging them in conservation efforts, we tap into their passion and commitment to safeguarding their natural environment. This grassroots involvement not only inspires a sense of ownership but also fosters a shared responsibility for the protection of our planet.

Local communities can contribute to conservation efforts in various ways. They can actively participate in habitat restoration projects, such as reforestation or wetland rehabilitation, thereby directly enhancing the quality of their ecosystem. By educating themselves and others about sustainable practices, they can promote environmentally friendly behaviors within their communities, ensuring the long-term protection of natural resources.

Additionally, local communities can collaborate with environmental engineering professionals to conduct research and gather data on their ecosystems. This valuable information can help identify the most effective conservation strategies and guide decision-making processes. By involving local communities, we ensure that conservation efforts are tailored to the specific needs and challenges of each region, leading to more successful and sustainable outcomes.

In conclusion, the active involvement of local communities is essential for achieving effective habitat conservation and restoration. By recognizing their invaluable knowledge, passion, and commitment, we can harness their potential to protect and restore our natural resources. Environmental engineering professionals, along with everyone interested in conservation, must embrace the role of local communities and actively engage them in our collective efforts towards a sustainable future. Together, we can create a blueprint for habitat conservation and restoration that is inclusive, effective, and long-lasting.

Collaborating with Government Agencies and NGOs

In the pursuit of a sustainable future, collaboration between various stakeholders is crucial. Government agencies and non-governmental organizations (NGOs) play a pivotal role in driving environmental engineering initiatives that aim to conserve and restore habitats. This subchapter explores the importance and benefits of collaborating with these entities, highlighting how their collective efforts can lead to significant positive change.

Government agencies possess the authority, resources, and regulatory power necessary to enact policies and regulations that promote habitat conservation and restoration. Their involvement ensures that environmental engineering projects comply with legal requirements, including environmental impact assessments, permits, and licensing. By collaborating with government agencies, environmental engineers can access valuable data, expertise, and funding opportunities, which can enhance the success and impact of their initiatives.

NGOs, on the other hand, often possess specialized knowledge, technical expertise, and a deep commitment to environmental conservation. These organizations are driven by a desire to protect and restore habitats, and they mobilize resources, raise awareness, and advocate for change. NGOs can provide environmental engineers with grassroots support, community engagement opportunities, and access to networks of like-minded individuals and organizations. Collaborating with NGOs allows environmental engineers to tap into their passion, expertise, and on-the-ground experience, which can greatly strengthen the effectiveness of habitat conservation and restoration efforts.

The benefits of collaborating with government agencies and NGOs extend beyond their respective resources and expertise. Such partnerships create a platform for knowledge sharing, fostering a cross-pollination of ideas, innovation, and best practices. By working together, environmental engineers, government agencies, and NGOs can combine their strengths, share data, and streamline their efforts, resulting in more efficient and impactful projects.

Furthermore, collaboration with government agencies and NGOs can also serve as a catalyst for public engagement and education. These entities often have established networks, platforms, and communication channels that can help raise awareness about the importance of habitat conservation and restoration. By involving the public in these initiatives, environmental engineers can foster a sense of ownership and responsibility, encouraging individuals to take action and make sustainable choices in their own lives.

In conclusion, collaborating with government agencies and NGOs is paramount for environmental engineers in their quest for a sustainable future. By leveraging the resources, expertise, and networks of these entities, environmental engineers can enhance the success and impact of their habitat conservation and restoration projects. Together, they can drive positive change, protect ecosystems, and create a sustainable future for all.

Educating and Empowering Individuals for Sustainable Action

In today's rapidly changing world, the need for sustainable action and environmental stewardship has become more crucial than ever. The challenges we face, such as climate change, habitat destruction, and resource depletion, require urgent attention and collective efforts from all individuals. This subchapter, "Educating and Empowering Individuals for Sustainable Action," aims to inspire and inform readers from all walks of life, including those interested in the field of environmental engineering, on how they can contribute to a sustainable future.

Education plays a pivotal role in shaping individuals' perspectives and actions towards the environment. By providing comprehensive and accessible information, we can empower people to make informed decisions and take sustainable actions in their everyday lives. This subchapter delves into the importance of environmental education, highlighting its role in creating awareness, fostering a sense of responsibility, and cultivating a mindset of sustainability. It explores various educational approaches and tools that can be utilized to engage individuals in environmental issues, including formal education systems, community-based programs, and digital platforms.

Furthermore, this subchapter emphasizes the significance of empowering individuals to take action. It highlights the transformative potential of small-scale actions and encourages readers to embrace their agency in creating positive change. By showcasing inspiring examples of individuals who have made a difference, it aims to demonstrate that sustainable action is not limited to experts or professionals but is accessible to everyone.

For those interested in the field of environmental engineering, this subchapter provides insights into the specific role they can play in promoting sustainability. It explores the intersection of engineering and environmental conservation, emphasizing the importance of incorporating sustainable practices, innovative technologies, and ethical considerations into engineering projects. It also discusses the vital role of collaboration between engineers, policymakers, and communities in developing sustainable solutions.

Overall, "Educating and Empowering Individuals for Sustainable Action" calls upon every individual to take responsibility for the future of our planet. It provides practical tips, resources, and case studies to guide readers in their journey towards sustainability. Whether you are an environmental engineering enthusiast or simply passionate about making a difference, this subchapter aims to inspire and equip you with the knowledge and tools necessary to contribute to a sustainable future. Together, we can create a blueprint for habitat conservation and restoration that ensures a thriving planet for generations to come.

Chapter 8: Case Studies in Successful Habitat Conservation and Restoration

Restoring Degraded Farmlands into Thriving Wildlife Habitats

In today's rapidly changing world, the need for sustainable practices and habitat conservation has become more critical than ever. One area that requires immediate attention is the restoration of degraded farmlands into thriving wildlife habitats. This subchapter explores the challenges faced by environmental engineers in this field and offers a blueprint for achieving a sustainable future.

Farmlands, which were once teeming with diverse ecosystems, have often been subjected to intensive agricultural practices resulting in degradation of soil, water, and biodiversity. The loss of natural habitats has had a significant impact on wildlife populations, leading to the decline of numerous species. It is the responsibility of environmental engineers to find innovative solutions that restore these degraded lands and create thriving habitats for wildlife.

One approach to restoring degraded farmlands is through the implementation of agroforestry systems. By integrating trees and shrubs into agricultural landscapes, engineers can improve soil health, prevent erosion, and provide a multitude of benefits for wildlife. Agroforestry systems offer nesting sites, food sources, and shelter for various species, ultimately creating a balanced ecosystem that supports biodiversity.

Furthermore, the implementation of conservation agriculture techniques is crucial for restoring degraded farmlands. By minimizing soil disturbance, using cover crops, and practicing crop rotation, environmental engineers can improve soil fertility, increase water infiltration, and enhance biodiversity. These practices not only restore the natural balance of the ecosystem but also contribute to the long-term sustainability of agricultural production.

To achieve successful restoration, it is essential to engage local communities and farmers in the process. Environmental engineers must collaborate with agricultural stakeholders, providing them with the necessary knowledge and resources to adopt sustainable practices. By fostering partnerships and promoting community involvement, the restoration of degraded farmlands can become a collective effort, ensuring a sustainable future for both wildlife and agricultural productivity.

In conclusion, the restoration of degraded farmlands into thriving wildlife habitats is a pressing issue that requires the expertise and dedication of environmental engineers. Through the implementation of agroforestry systems and conservation agriculture techniques, these professionals can create sustainable solutions that promote biodiversity and restore the natural balance of ecosystems. By involving local communities and farmers in the restoration process, we can ensure a brighter future where farmlands coexist harmoniously with thriving wildlife habitats. Together, we can build a sustainable future for the benefit of every living being on the planet.

Urban Greening Initiatives: Transforming Cities into Biodiversity Hotspots

In recent years, the rapid urbanization and expansion of cities have led to significant challenges for both humans and the environment. The concrete jungles that cities have become are often devoid of green spaces, resulting in a multitude of environmental issues such as air pollution, heat island effect, and the loss of biodiversity. However, there is hope on the horizon as urban greening initiatives are gaining momentum worldwide, aiming to transform cities into biodiversity hotspots while addressing these pressing environmental concerns.

Urban greening initiatives encompass a range of strategies and practices that aim to reintroduce nature into cities, making them more sustainable, resilient, and livable for both humans and wildlife. These initiatives include the creation of green roofs and walls, urban forests, community gardens, and the preservation and restoration of existing natural areas within cities. They are not only transforming the physical landscape of cities but also fostering a sense of community and connection with nature among urban dwellers.

One of the key benefits of urban greening initiatives is the improvement of air quality. Trees and plants play a vital role in absorbing harmful pollutants and carbon dioxide while releasing oxygen, resulting in cleaner and healthier air for city residents. Moreover, the presence of green spaces in cities helps to mitigate the urban heat island effect, reducing the overall temperature and providing much-needed shade and cooling in hot summer months.

Beyond the environmental benefits, urban greening initiatives also contribute to the conservation and restoration of biodiversity. By creating green corridors and connecting fragmented habitats, these initiatives create vital pathways for wildlife, allowing them to move freely within cities and promoting species diversity. Urban green spaces become havens for birds, insects, and other wildlife, providing them with food, shelter, and breeding grounds that are often lacking in urbanized areas.

For the audience of environmental engineering professionals, urban greening initiatives present a unique opportunity to apply their expertise in designing and implementing sustainable solutions for cities. From developing innovative green infrastructure to optimizing water and energy use, environmental engineers can play a crucial role in shaping the future of urban greening initiatives.

In conclusion, urban greening initiatives are transforming cities into biodiversity hotspots, providing numerous benefits for both the environment and human well-being. These initiatives not only improve air quality, reduce the urban heat island effect, and enhance the overall livability of cities but also contribute to the conservation and restoration of biodiversity. As environmental engineering professionals, it is our responsibility to embrace and promote these initiatives, working towards a sustainable future where cities coexist harmoniously with nature.

Mangrove Restoration: Protecting Coastal Communities and Marine Life

Mangroves are incredible ecosystems that hold immense value for both coastal communities and marine life. In recent years, the degradation and loss of mangrove forests have become a growing concern for environmental engineers and conservationists alike. The subchapter "Mangrove Restoration: Protecting Coastal Communities and Marine Life" explores the importance of mangroves, the threats they face, and the strategies for their restoration.

Mangroves, found in coastal areas of tropical and subtropical regions, are highly productive and biodiverse habitats. They serve as a crucial buffer zone between land and sea, protecting coastlines from erosion and storm surges. These unique trees have adapted to thrive in saline conditions, with their extensive root systems providing stability and acting as nurseries for a wide range of marine species. Additionally, mangroves act as carbon sinks, sequestering significant amounts of atmospheric carbon dioxide and mitigating climate change impacts.

Despite their ecological significance, mangrove forests are under severe threat due to human activities such as deforestation, urbanization, and aquaculture. The consequences of mangrove loss are far-reaching, affecting both the environment and local communities. Coastal erosion increases, leaving vulnerable communities exposed to the destructive force of storms and tsunamis. The loss of habitat disrupts the delicate balance of marine ecosystems, impacting fish populations and reducing the resilience of coastal fisheries.

Fortunately, environmental engineers and conservationists have recognized the urgent need for mangrove restoration. The subchapter outlines various restoration techniques, including reforestation, hydrological restoration, and community-based initiatives. Reforestation involves planting mangrove saplings in areas where they have been lost or degraded. Hydrological restoration focuses on reinstating natural water flows in degraded mangrove areas, facilitating their recovery. Engaging local communities in restoration efforts is crucial, as their livelihoods often depend on the health of the mangroves.

Furthermore, the subchapter emphasizes the importance of integrating mangrove restoration with sustainable development goals. By promoting alternative livelihood options for local communities, such as eco-tourism or sustainable aquaculture, the pressure on mangrove resources can be reduced. Implementing effective monitoring and management strategies is also vital to ensure the long-term success of restoration projects.

In conclusion, the subchapter "Mangrove Restoration: Protecting Coastal Communities and Marine Life" highlights the significance of mangroves in coastal ecosystems and the urgent need for their restoration. It provides insights into the threats faced by mangroves, the restoration techniques available, and the importance of community involvement. By restoring and protecting mangrove forests, environmental engineers and conservationists can contribute to a sustainable future, safeguarding both coastal communities and marine life.

Chapter 9: Challenges and Future Directions

Overcoming Financial and Political Barriers

In the pursuit of a sustainable future, one of the biggest challenges we face is the presence of financial and political barriers. These obstacles often hinder progress in environmental engineering and habitat conservation and restoration efforts. However, it is crucial to understand that these barriers can be overcome with the right strategies and collective action.

Financial barriers are perhaps the most common roadblock encountered in the field of environmental engineering. The cost of implementing sustainable practices, developing conservation projects, and restoring habitats can be substantial. Many individuals and organizations find it difficult to secure adequate funding to support their initiatives. However, it is important to recognize that investing in sustainable solutions can lead to long-term cost savings and environmental benefits.

To overcome financial barriers, various approaches can be adopted. One effective strategy is to foster partnerships between public and private sectors. Collaboration between government agencies, businesses, and non-profit organizations can pool resources and share the financial burden. Additionally, seeking grants and funding opportunities specifically designed for environmental projects can provide the necessary financial support.

Political barriers, on the other hand, often arise due to conflicting interests and lack of political will. Environmental issues can be polarizing, and it is not uncommon for politicians to prioritize short-term gains over long-term sustainability. Overcoming political barriers requires a multi-faceted approach that includes raising awareness, advocating for policy changes, and engaging with decision-makers.

Education and awareness play a vital role in overcoming political barriers. By educating the public and decision-makers about the importance of habitat conservation and restoration, we can build a broader consensus for environmental action. This can be done through workshops, conferences, and public campaigns aimed at highlighting the benefits of sustainable practices.

Advocacy and lobbying efforts are also crucial in driving policy changes. Engaging with local, regional, and national policymakers can help redirect political focus towards sustainable development. By presenting evidence-based arguments and showcasing successful case studies, environmental engineers and conservationists can influence decision-making processes and secure political support.

In conclusion, while financial and political barriers pose significant challenges in the pursuit of a sustainable future, they are not insurmountable. Through collaboration, innovative funding mechanisms, education, and advocacy, we can overcome these barriers and pave the way for effective habitat conservation and restoration. By addressing these obstacles head-on, we can ensure a sustainable future for ourselves, future generations, and the planet as a whole.

Integrating Conservation and Restoration Efforts on a Global Scale

In today's rapidly changing world, the need for conservation and restoration efforts has never been more pressing. Climate change, deforestation, and the loss of biodiversity pose significant threats to our planet and its delicate ecosystems. Addressing these challenges requires a coordinated and integrated approach that can be implemented on a global scale.

This subchapter explores the importance of integrating conservation and restoration efforts to achieve a sustainable future for all. It emphasizes the role of environmental engineering in developing innovative solutions that can help restore damaged habitats and protect vulnerable ecosystems.

Conservation and restoration are two interrelated strategies that aim to safeguard the natural environment. While conservation focuses on preserving existing habitats and species, restoration involves the rehabilitation and revitalization of degraded or destroyed ecosystems. By combining these approaches, we can create a comprehensive framework that addresses both the preservation of biodiversity and the recovery of damaged environments.

At a global level, integrating conservation and restoration efforts requires collaboration between governments, organizations, and individuals. By sharing knowledge, resources, and expertise, we can develop effective strategies that transcend national borders and tackle environmental challenges collectively.

Environmental engineering plays a crucial role in this process by providing the technical expertise needed to develop sustainable solutions. Through innovative technologies and practices, environmental engineers can design and implement restoration projects that promote biodiversity, enhance ecosystem services, and mitigate the impacts of climate change.

Furthermore, integrating conservation and restoration efforts on a global scale can yield numerous benefits. It can help preserve endangered species, protect vital ecosystems such as forests and wetlands, and maintain the delicate balance of our planet's natural systems. Additionally, these efforts can contribute to climate change mitigation by restoring carbon sinks and promoting sustainable land-use practices.

However, achieving successful integration requires overcoming various challenges, including political barriers, limited funding, and the need for long-term commitment. This subchapter explores these obstacles and proposes strategies to overcome them, emphasizing the importance of public awareness and education in driving positive change.

In conclusion, integrating conservation and restoration efforts on a global scale is essential for creating a sustainable future for all. By combining the principles of conservation and restoration with the expertise of environmental engineering, we can develop innovative solutions that protect our planet's biodiversity and restore its fragile ecosystems. Together, we can make a significant difference and ensure a brighter future for generations to come.

Embracing Technology and Innovation for Effective Habitat Management

In today's rapidly advancing world, the integration of technology and innovation has become crucial for effective habitat management. The Sustainable Future: A Blueprint for Habitat Conservation and Restoration aims to provide insights into how embracing these tools can contribute to a sustainable and thriving environment. This subchapter explores the benefits and opportunities that arise from utilizing technology and innovation in the field of environmental engineering.

Environmental engineering is a niche that plays a pivotal role in habitat conservation and restoration. It encompasses various disciplines such as water resource management, pollution control, and sustainable development. By incorporating technology and innovation, environmental engineers can significantly enhance their ability to address the challenges faced in habitat management.

One of the key benefits of embracing technology is the ability to collect and analyze vast amounts of data. With the advent of sensors, drones, and remote sensing technologies, environmental engineers can now monitor habitats in real-time and gather precise information about their condition. This data-driven approach allows for more informed decision-making and targeted interventions to conserve and restore habitats.

Furthermore, innovative technologies can revolutionize the way we approach habitat management. For instance, the utilization of artificial intelligence (AI) and machine learning algorithms can help identify patterns and predict habitat changes. This enables proactive measures to be taken, minimizing the negative impacts on ecosystems and biodiversity.

The integration of technology also promotes greater engagement and awareness among stakeholders. Through interactive platforms and mobile applications, the general public can actively participate in habitat management efforts. Citizen science initiatives, where individuals contribute data and observations, have proven to be valuable in monitoring and conserving habitats. This inclusive approach fosters a sense of ownership and responsibility, leading to a more sustainable future.

In addition to data collection and stakeholder engagement, technology and innovation also play a crucial role in developing sustainable solutions. Advanced techniques such as bioremediation, ecological engineering, and habitat restoration using 3D printing are all examples of innovative approaches that can help restore habitats and improve their resilience to climate change.

Embracing technology and innovation is no longer an option but a necessity for effective habitat management. This subchapter serves as a guide for environmental engineers, providing them with insights into the numerous benefits and opportunities that arise from integrating technology into their practices. By collectively harnessing the power of technology, we can ensure a sustainable future where habitats are conserved, restored, and flourish for generations to come.

Chapter 10: The Role of Every Individual in Creating a Sustainable Future

Practical Steps for Promoting Habitat Conservation in Daily Life

In our rapidly changing world, it has become more crucial than ever to take action and promote habitat conservation in our daily lives. The Sustainable Future: A Blueprint for Habitat Conservation and Restoration aims to provide a comprehensive guide on how individuals from all walks of life can contribute to this noble cause. This subchapter, "Practical Steps for Promoting Habitat Conservation in Daily Life," is specifically tailored for the audience of environmental engineering, but the strategies and tips provided can be implemented by anyone interested in making a positive impact on our planet.

1. Reduce, Reuse, Recycle: One of the most effective ways to promote habitat conservation is by adopting a sustainable lifestyle. Start by reducing your consumption, reusing items whenever possible, and recycling waste products. This simple practice helps conserve resources and reduces the strain on natural habitats.

2. Choose Sustainable Products: When purchasing goods, opt for products that are eco-friendly and have minimal impact on the environment. Look for labels like Fair Trade, Forest Stewardship Council (FSC), and Energy Star. By supporting sustainable businesses, you contribute to the conservation of habitats.

3. Conserve Water: Water scarcity is a global issue, and conserving water is vital for habitat preservation. Implement simple measures such as fixing leaky faucets, using low-flow showerheads, and collecting rainwater for gardening. These actions not only save water but also protect aquatic ecosystems.

4. Support Local Conservation Efforts: Get involved in local conservation organizations and initiatives. Volunteer your time, participate in habitat restoration projects, and spread awareness about the importance of preserving natural habitats. By working together, we can make a significant impact.

5. Plant Native Species: When landscaping or gardening, choose native plant species. Native plants provide food and shelter for local wildlife, promote biodiversity, and require fewer resources to maintain. Create a habitat-friendly environment in your backyard or community and encourage others to do the same.

6. Educate and Inspire: Share your knowledge and passion for habitat conservation with others. Use social media, blogs, or public speaking opportunities to raise awareness about the importance of preserving habitats and inspire others to take action. Education is a powerful tool for change.

7. Practice Responsible Tourism: When traveling, choose eco-friendly accommodations and activities that respect and support local ecosystems. Avoid activities that harm wildlife or disrupt natural habitats. By being mindful of our impact as tourists, we can promote sustainable tourism practices.

By implementing these practical steps in our daily lives, we can collectively contribute to habitat conservation and restoration. Remember, every action counts, and even small changes can make a significant difference. Let's work together to create a sustainable future for all living beings on our beautiful planet.

Inspiring Others to Join the Movement

The fight for a sustainable future and the conservation and restoration of our habitats is a task that cannot be accomplished alone. It requires the collective effort and collaboration of individuals from all walks of life. Whether you are an environmental engineer, a concerned citizen, or simply someone who cares about the well-being of our planet, we all have a role to play in creating a sustainable future.

Environmental engineering, as a specialized field, holds immense potential to drive the necessary change towards sustainability. By applying scientific and engineering principles, environmental engineers can develop innovative solutions to address environmental challenges. However, their impact can be amplified by inspiring others to join the movement and work towards a shared vision.

One of the most effective ways to inspire others is through education and awareness. By sharing knowledge about the importance of habitat conservation and restoration, we can empower individuals with the understanding and motivation to take action. This can be achieved through workshops, training programs, and public lectures that highlight the urgent need for sustainable practices and their potential benefits.

Another powerful tool is storytelling. Humans are naturally drawn to narratives that evoke emotions and create connections. By sharing stories of successful habitat conservation and restoration projects, we can inspire others by showcasing the positive impact that can be achieved. These stories can be shared through books, documentaries, social media platforms, and even personal interactions.

Leadership also plays a crucial role in inspiring others. Environmental engineers have the opportunity to take on leadership roles and act as catalysts for change. By leading by example and demonstrating their commitment to sustainability, they can inspire others to follow suit. This can be done through initiating projects, organizing community clean-ups, or actively participating in environmental advocacy.

Lastly, collaboration and partnerships are essential in inspiring others to join the movement. By working together with individuals and organizations from different sectors, we can create a united front in the fight for a sustainable future. Environmental engineers can collaborate with scientists, policymakers, businesses, and community groups to develop comprehensive solutions that address the complex challenges we face.

In conclusion, inspiring others to join the movement towards a sustainable future is a critical step in habitat conservation and restoration. Environmental engineers, as leaders in this field, have the power to educate, tell stories, lead by example, and collaborate with others to inspire change. By engaging a diverse audience and encouraging their participation, we can create a collective voice that drives the transformation needed for a sustainable planet.

Building a Network of Sustainable Communities

In today's rapidly changing world, the need for sustainable communities has become more pressing than ever before. As the effects of climate change continue to intensify, it is crucial for us to build a network of sustainable communities that can withstand environmental challenges and provide a high quality of life for their residents. This subchapter will explore the key principles and strategies for creating such communities, with a focus on the role of environmental engineering.

Environmental engineering plays a vital role in designing and implementing sustainable solutions for communities. With their expertise, environmental engineers can develop innovative technologies and practices that minimize environmental impacts and promote resource efficiency. By integrating sustainable infrastructure, such as renewable energy systems, green buildings, and water conservation measures, environmental engineers help create communities that are self-sufficient and resilient.

One of the fundamental principles of building sustainable communities is the concept of smart growth. This involves designing communities in a way that minimizes sprawl, reduces transportation needs, and promotes walkability and public transit. Environmental engineers can contribute to smart growth by designing efficient transportation systems, promoting the use of electric vehicles, and implementing green spaces that encourage physical activity and social connectivity.

Another crucial aspect of sustainable communities is the management of natural resources. Environmental engineers can play a key role in developing sustainable waste management systems, recycling programs, and water conservation strategies. By implementing innovative technologies, such as anaerobic digestion for organic waste or rainwater harvesting systems, communities can reduce their environmental impact and preserve valuable resources for future generations.

Moreover, building a network of sustainable communities requires a holistic approach to planning and decision-making. Environmental engineers can contribute to this process by conducting comprehensive environmental impact assessments, considering the long-term implications of development projects, and involving stakeholders in the decision-making process. By incorporating sustainability principles into the planning stage, communities can avoid costly and environmentally damaging retrofits in the future.

The benefits of building a network of sustainable communities are numerous. Not only do they provide a higher quality of life for their residents, but they also contribute to the overall well-being of the planet. By reducing greenhouse gas emissions, conserving resources, and preserving natural habitats, these communities serve as models for a sustainable future.

In conclusion, building a network of sustainable communities is a critical step towards creating a future that is environmentally responsible and resilient. Environmental engineering plays a crucial role in designing and implementing sustainable solutions for communities. By adopting principles such as smart growth, resource management, and holistic planning, we can create communities that thrive in harmony with nature. Together, we can build a sustainable future for all.

Conclusion: A Call to Action for a Sustainable Future

As we reach the end of this book, "The Sustainable Future: A Blueprint for Habitat Conservation and Restoration," it is essential to emphasize the urgency and importance of taking action in achieving a sustainable future. This conclusion serves as a call to action not only to environmental engineers but to everyone committed to making a positive impact on our planet.

The world is facing unprecedented environmental challenges, from climate change to habitat destruction, biodiversity loss, and resource depletion. It is no longer enough to simply acknowledge these issues; we must act collectively and decisively to reverse the damage and build a sustainable future for generations to come.

Environmental engineers play a crucial role in this endeavor. Their expertise in designing, developing, and implementing sustainable solutions is vital for addressing the complex environmental problems we face. From renewable energy systems to waste management strategies, environmental engineers have the power to shape a more sustainable future.

However, the responsibility for creating change does not solely rest on the shoulders of environmental engineers. The call to action extends to every individual, regardless of their profession or background. We must all recognize our interconnectedness with the natural world and take responsibility for our actions.

The first step in creating a sustainable future is education and awareness. We must educate ourselves and others about the environmental challenges we face and the potential solutions available.

By understanding the impact of our actions, we can make informed decisions and change our behaviors accordingly.

Secondly, we must support and advocate for policies and initiatives that promote sustainability. This includes advocating for renewable energy, sustainable agriculture, and responsible consumption. By using our voices and voting power, we can influence policy decisions that prioritize the long-term health of our planet.

Finally, we must take practical actions in our everyday lives. This can involve reducing our carbon footprint, conserving water, recycling, supporting local and sustainable businesses, and protecting natural habitats. Small actions can have a ripple effect, inspiring others to do the same and creating a collective impact.

In conclusion, the time for action is now. "The Sustainable Future: A Blueprint for Habitat Conservation and Restoration" serves as a guide and a source of inspiration, but it is up to every individual, including environmental engineers and every member of society, to take action. By working together, we can forge a sustainable future for ourselves, future generations, and the planet we call home.